YOUR KNOWLEDGE HAS VALUE

- We will publish your bachelor's and
  master's thesis, essays and papers

- Your own eBook and book -
  sold worldwide in all relevant shops

- Earn money with each sale

Upload your text at www.GRIN.com
and publish for free

Imprint:

Copyright © 2018 GRIN Verlag
Print and binding: Books on Demand GmbH, Norderstedt Germany
ISBN: 9783346004840

This book at GRIN:

https://www.grin.com/document/490589

Kedija Hussen Mohammed

Milk adulteration. Options to maintain a quality product

GRIN Verlag

MILK ADULTERATION: OPTIONS TO MAINTAIN QUALITY PRODUCT

BY

KEDIJA HUSSEN

REVIEW PAPER

FEBRUARY, 2018

BISHOFTU, ETHIOPIA

TABLE OF CONTENTS   **PAGE**

1. **INTRODUCTION** 1

2. **MILK ADULTERATION** 2

    2.1   Milk Adulterants 3

        2.1.1. Qualitative (Biochemical) detection methods 8

        2.1.2 Quantitative (Biotechnological) detection methods 13

3. **MILK ADULTERANTS AND RELATED HEALTH COMPLICATIONS** 15

4. **CAUSE FOR MILK ADULTERATION** 16

5. **OPTIONS TO MINIMIZE ADULTRATED MILK PRODUCTS** 16

    5.1 Use of easy milk tester at the spot 16

        5.1.1. Machine detection methods 16

        5.1.2. Ready to use kit 17

    5.2. IMPROVE PROFIT MARGIN OF DAIRY FARMERS 17

    5.2.1. Improving Milk Production by Biotechnological Application 18

        5.2.1.1 Improving milk production through reproductive biotechnology 18

        5.2.1.2 Improve milk quality through processing technology 20

        5.2.2 Labeling of Milk quality (*as A1 and A2* milk) 21

6. **CONCLUSIONS AND RECOMMENDATIONS** 25

7. **REFERENCES** 27

1. INTRODUCTION

Milk is an important component of diets for all humans as it is high in essential amino acids that are most likely to be deficient in diets based on vegetable protein. Milk is considered to be the 'ideal food' because of its abundant nutrients required by both infants and adults (Faraze *et al.*, 2013; Renny *et al.*, 2014). Milk contains 87% water, 3.3% protein, 3.9% fats, 5% lactose and 0.7% ash. In addition, it is a raw component to different food products. For example, casein protein can be separated and then broken down to provide a nitrogen source for infant formula and special diets for critically ill people. In addition, it is one of the best sources for protein, fat, carbohydrate, vitamin and minerals (Faraz *et al.*., 2013). Hence, milk is a highly nutritional natural product that can be used for a wide range of different products.

Although milk is a high-cost source of protein and fat relative to vegetable sources, it is readily saleable particularly in the more affluent urban areas of developing countries. Improving milk production is therefore an important tool for improving the quality of life particularly for rural people in developing countries. However, the production level of dairy animal is not satisfying the consumer demand worldwide. This became one of the driving forces for milk adulteration by the producer (Moore *et al.*, 2012). Possible reasons behind it may not only include demand and supply gap, but also perishable nature of milk, low purchasing capability of customer and lack of suitable detection methods (Kamthania *et al.*, 2014) also aggravate the problem. The motivation for food fraud is economic, but the impact is a real public health concern (Ellis *et al.*, 2012; Singh & Gandhi, 2015; Moore *et al.*, 2012).

Milk and dairy product adulteration came into global concern after breakthrough of melamine contamination in Chinese infant milk products in 2008. However, history of milk adulteration is very old. Swill milk scandal has been reported in 1850 which killed 8000 infants in New York alone (Xin & Stone, 2008).

Milk adulteration is significantly worse in developing and underdeveloped countries due to absence of adequate monitoring and lack of proper enforcement policies (Xin and Stone, 2008). However, this is one of the most common phenomena that have been overlooked in many countries (Tanzina and Shoeb, 2016). Hence, this paper reviews current milk adulteration possibilities, related public health risks and available technologies options in controlling milk quality through different detection methods.

2. MILK ADULTERATION

Food adulteration was first discovered in 1820 by the German chemist Friedrich Accum. He identified many toxic metal in food and drinks. Frederick Accum was the first to raise the alarm about food adulteration. By that time Accum had become aware of the problem through his analytical work and this led him to publish 'A treatise on adulterations of food and culinary poisons '. It was the first serious attempt to expose the nature, extent and dangers of food adulteration. In his book cheese was one of the product that found to be adulterated with red lead (Pb_3O_4 C and vermilion (mercury sulphide, HgS) for coloring. Since then milk adulteration continues to be a serious global issue (Noel, 2005). Based on different review works undertaken by Moore *et al.* (2012) indicated that milk is the second most likely food item being in the risk of adulteration after olive oil.

Definition of adulteration of food: Food adulteration is the process in which the quality of food is lowered either by the addition of inferior quality material or by extraction of valuable ingredient. This process substitutes the main product wholly or partly and the added substances may be injurious to health (Spink and Moyer, 2011).

Milk Adulteration is defined as the removal or replacement of milk components with inferior substances without a purchaser's knowledge (Bansal and Singhal, 1991; Santos *et al.*, 2013). Hence, based on the above definitions, a food item is said to be adulterated if:
- A valuable ingredient has been abstracted from the food product, wholly or in part.
- An inferior substance substitutes wholly or partly
- A substance which is added is injurious for human consumption.

2.1 Milk Adulterants

Various types of adulterants found in the food products can be classified in to intentional and incidental adulterants (like pesticide residues, larvae in foods, droppings of rodents). But this review paper discuss only about intentional adulterants in milk.

It is a well-established fact that consumers want clean, wholesome and nutritious milk that is produced and processed in a sound, sanitary manner and is free from pathogens and foreign substances. Hence, for fulfilling consumer's demand, quality milk production is necessary. Quality dairy products means, the products which are free from pathogenic bacteria, harmful toxic substances, and sediment and extraneous substances, and have good flavor, low in bacterial counts, with normal composition (Khan *et al.*, 2008).

The industries usually define milk quality based on the nutrient levels, mainly protein and fat besides of its pathogenic quality. These parameters have been used to calculate the payout to the supplier (Embrapa, 2014). Generally, common parameters that are checked to evaluate milk quality are fat percentage, protein percentage, SNF (Solid-not-Fat) percentage, and freezing point (Tanzina and Shoeb, 2016). Adulterants are added in milk to increase these parameters, thereby increasing the milk quality in dishonest way (Moore *et al.*, 2012).

Hence, deliberately dairy farmers or traders adulterate the milk with different products. Commonly, water is added to milk to increase volume which decreases specific gravity of milk. Then to retain the gravity, shelf life, and color of the milk hazardous chemicals are added (Kumar *et al.*, 1998; Fischer et al., 2011; Arvind *et al.*, 2012; Sharma *et al.*, 2012a).

Adulterants in milk mainly include addition of vegetable protein, milk from different species, addition of whey and watering which are known as economically motivated adulteration (Fischer *et al.*, 2011; Singh & Gandhi, 2015). These adulterations do not pose any severe health risk. However, some adulterants are too harmful to be overlooked. Some of the major adulterants in milk having serious adverse health effect are urea, formalin, detergents, ammonium sulphate, boric acid, caustic soda, benzoic acid, salicylic acid, hydrogen peroxide, sugars and melamine

(Kumar *et al.*, 1998; Moore *et al.*, 2012). Some of the common adulterants found in milk and their detection methods are discussed bellow.

1) water

Water is most common, oldest and simplest methods of milk adultrant, which is added in to milk for economic purpose. Since it artificially increases its volume for greater profit; this can substantially decrease the nutritional value of milk. If the water added is contaminated, there is a of potential waterborne diseases (Kandpal *et al.*, 2012). Then to retain the shelf life, and to compensate for density and color, milk is adulterated with harmful chemicals thereby to pass quality test during milk collection.

2) Sugar

Table sugar is added to milk to increase the carbohydrate content of the milk and thus the density (solids not fat content) of milk will be increased and it make the milk thicker. Generally, it increases the lactometer reading of milk, which was already diluted with water (Sharma *et al.*, 2012a; Kamthania *et al.*, 2014; Arvind *et al.*, 2012).

3) Starch

Starch is added to milk to increases milk solid or SNF content of milk. (Sharma *et al.*, 2012a; Arvind et al.., 2012; Kumar *et al.*, 1998). Apart from the starch, wheat flour, and rice flour are also added (Sharma *et al.*, 2012a).

4) Acids

Acids like Benzoic acid and Salicylic acid is used as a preservative in food industry. It also added to milk to preserve and thus increase the shelf life of milk (Arvind *et al.*, 2012).

5) Soap/detergent

Detergents are added to emulsify and dissolve the oil in water giving a frothy solution, which is the desired characteristics of milk (Singuluri & Sukumaran, 2014) and thus to have thick milk. Addition of such chemicals will cause health problem especially related to stomach and kidneys.

6) Formalin

Formalin is added as a preservative and can preserve milk for long period of time (Singh & Gandhi, 2015); (Arvind *et al.*, 2012); (Kamthania *et al.*, 2014). Due to its high toxicity, it is considered to cause liver and kidney damage.

7) Urea in milk

Urea is added to milk as a preservative. Urea is generally added in the preparation of synthetic milk to raise the SNF value and increase non-protein nitrogen content (Sharma *et al.*, 2012a). It also adds viscosity to milk thereby giving a feeling of thick milk.

8) Neutralizers in milk

Neutralizers are added to neutralize the acidic sour milk and help to avoid spoilage of milk. Common neutralizers which are added to milk are hydrated lime, sodium hydroxide, baking soda or sodium bicarbonate and washing soda or sodium carbonate (Sharma *et al.*, 2012a; Sowmya et al., 2013).

9) Sodium chloride/salt in milk

Salt in milk is mainly added with the aim of increasing the corrected lactometer reading by increasing the density of milk (Sharma *et al.*, 2012a).

10) Ammonium Sulphate

Ammonium Sulphate is added to the milk to increases the lactometer reading by maintaining the density of diluted milk (Singh & Gandhi, 2015).

11) Hydrogen peroxide

It is added to milk as preservative and then to increase the shelf life of milk (Singh & Gandhi, 2015); (Arvind *et al.*, 2012); (Kamthania *et al.*, 2014); (Sharma *et al.*, 2012a). However, it is hazardous chemical for internal use and it cause for kidney failure.

12. Skim milk powder in milk

Powder milk is added to fresh milk in the aim of producing synthetic milk. May be the powder milk is expired.

13. Vegetable oil in milk

Milk contains relatively large amount of fat and milk fat is very expensive. Hence, by extracting valuable component of natural fat present in milk, the chip vegetable oil is replaced to milk. Fat is removed as butter or cream (Willibald *et al.*, 2002; Singuluri & Sukumaran, 2014).

14) Milk adulteration with melamine

Price of milk is set based on ingredient quality (nutritional parameters) and these have been used to calculate the payout to the supplier (Embrapa, 2014). Since, companies using the milk for further production (e.g. of powdered infant formula) normally milk is checked the protein level through a test measuring nitrogen content (Liu *et al.*, 2012). These industry standard tests for protein content measure nitrogen and correlate the result to protein. This means that adding high nitrogen content compounds to protein-containing foods generates a higher protein content reading and hence a higher price. To this end the addition of cheap ingredients such as melamine increases the nitrogen content of the milk in order to pass a protein test (Liu *et al.*, 2012).

The addition of melamine was known after the 2008 milk adulteration scandal in China. As a result, in China, more than 50,000 infants have been hospitalized and 6 persons of them died because of high level of melamine in their food. Expert committees from World Health Organization (WHO) and Food and Drug Administration (FDA) investigators revealed that the cause of deaths was due to the crystal formation and blockage of renal tubes leading to failure in the functioning of kidney (Chikkappa, 2014; Maryam, 2017).

Melamine is an organic base chemical most commonly found in the form of white crystals and it is rich in nitrogen. Melamine is widely used in plastics, adhesives, countertops, dishware, and white boards.

Since melamine is neither a permitted additive nor a food ingredient, its limit had not been set in food legislation until the melamine contamination reported in China. Both the European Commission and the United States Food and Drug Administration (USFDA) have applied a maximum acceptable limit of 2.5 mg/kg for melamine in imported foods, and 1 mg/kg in infant formula (FAO, 2010; Liu *et al.*, 2010; Domingo *et al.*, 2014).

Melamine is not only added to milk as adulterant, but also in many other foods like brand yogurt, canned coffee, wheat gluten, chicken feed, and (Ingelfinger, 2008; Lin *et al.*, 2008). All these products were most probably manufactured using ingredients made from melamine-contaminated milk (PerkinElmer, 2013). It is also found in processed animal foods. In 2007, melamine was found in wheat gluten and rice protein concentrate exported from China and used in the manufacture of pet food in the United States. This caused the death of a large number of dogs and cats due to kidney failure (Cheng *et al.*, 2010; PerkinElmer, 2013; Maryam, 2017).

2.2. Methods of Detecting Adulterants in Milk

Quality control tests for milk are very important to assure adulterant free milk for consumption. Adulteration of milk reduces the quality of milk and can even make it hazardous. In general, these classified in to qualitative and quantitative detection methods. Qualitative detection methods are color based chemical reactions (biochemical) while quantitative detection methods are complex and diverse (Tanzina and Shoeb, 2016). This includes biotechnological and electrical methods. However, qualitative detections are advantageous because these are simple, rapid and very easy to perform (Tanzina and Shoeb, 2016).

2.1.1. Qualitative (Biochemical) detection methods

1) Detection of Water

Adulteration of milk with water can be checked by easy and cheap method of lactometer reading. However, addition of other adulterants affects the lactometer reading. At this case, hence freezing point of milk is an important indicator of the milk quality. The freezing point of milk is determined primarily to prove milk adulteration with water and to determine the amount of water in it. The presence of any substance can influence freezing point of product (Ilie *et al.*, 2010). Normal freezing point of milk is taken as –0.52°C- (-0.55°C). However, adulterated milk by water raises the freezing point towards 0°C. The presence of any solutes will depress freezing point below zero degrees C. Freezing point of milk is influenced by factors related to variation in environment, management, and breed (Henno *et al.*, 2008). Hence, the set off point for normal freezing point should be set on national level. Different country has different set point or standard for milk freezing point.

2) Detection of Table sugar
3)

Sugar will react with the resorcinol to give a red colored precipitate. It indicates the presence of table sugar in milk (Sharma *et al.*, 2012a); (Kamthania *et al.*, 2014); (Arvind *et al.*, 2012). Take 10 ml of milk in a test tube and add 5 ml of hydrochloric acid along with 0.1 g of resorcinol. Then shake the test tube well and place the test tube in a boiling water bath for 5 min. Appearance of red colour indicates the presence of added sugar in milk.

4) Detection of Starch

The test to detect starch in milk uses iodine solution, addition of which turns the milk solution to blue black color due to the formation of starch-iodine complex. Measure out 3ml milk into a test tube. The test tube is then kept for incubation in boiling water bath for 5 minutes. Then milk is cooled to room temperature and added with 2 to 3 drops of 1% iodine solution. Appearance of blue black color indicates that the milk is *adulterated* with starch (Sharma *et al.*, 2012a).

5) Detection of Acids

Presence of these acids can be detected by adding concentrated sulphuric acid and ferric chloride, which when reacts with benzoic acid and salicylic acid to give buff colored and violet colored reaction products.

Measure out 5ml milk into a test tube. Add few drops of concentrated sulphuric acid and gently shake the test tube. Add 0.5% ferric chloride solution drop wise into the test tube. Mix the contents well. Development of buff color is indicative of the presence of benzoic acid and if violet color is observed shows the presence of salicylic acid.

5) Detection of Soap and detergents in milk

A. Test for detection of soap

Take 10 ml of milk in a test tube and dilute it with equal quantity of hot water and then add 1 – 2 drops of phenolphthalein indicator. Development of pink colour indicates that the milk is adulterated with soap (Arvind *et al.*, 2012); (Kamthania *et al.*, 2014)

B. Test for Detection of detergents

Take 5 ml of milk in a test tube and add 0.1 ml of bromocresol purple solution. Appearance of violet colour indicates the presence of detergent in milk. Unadulterated milk samples show a faint violet colour (Singhal, 1980); (Arvind *et al.*, 2012).

6) Detection of Formalin

Formalin reacts with Sulphuric acid and ferric chloride to give a purple colored ring at the junction of the milk layers, thereby indicating the presence of formalin adulterated in milk (Arvind *et al.*, 2012); (Kamthania *et al.*, 2014)

A. Measure out 2ml milk into a test tube. Gently add 2ml of 90% Sulphuric acid and ferric chloride mixture into the test tube. Formation of purple color ring at the interface of two layers indicates that the milk is adulterated with formalin.

B. Take 10 ml of milk in test tube and 5 ml of conc. sulphuric acid with a little amount of ferric acid is added on the sides of the test tube without shaking. If a violet or blue ring appears at the intersection of the two layers, then it shows the presence of formalin.

7) Detection of urea

1. Five ml of milk is mixed well with 5 ml paradimethyl amino benzaldehyde (16%). If the solution turns yellow in colour, then it indicates the sample of milk is adulterated with urea (Sharma *et al.*, 1993); (Arvind *et al.*, 2012),

2. Take 5 ml of milk in a test tube and add 0.2 ml of urease (20 mg / ml). Shake well at room temperature and then add 0.1 ml of bromothymol blue solution (0.5%). Appearance of blue colour after 10-15 min indicates the adulterated milk with urea (Sharma *et al.*, 1993); (Arvind *et al.*, 2012).

8) Detection of neutralizer

Freshly drawn milk has an acidity of 0.12-0.16 per cent expressed as lactic acid. With the passage of time the acidity increases during souring of milk. Any acidity above 0.18 per cent lactic acid coagulates milk. This is due to the formation of lactic acid from lactose. Hence, milk producers and farmers tend to neutralize the milk to avoid rejection of milk with increased shelf-life at the milk collection centers and at the dairy plants. The common neutralizers which are added to milk are caustic or sodium hydroxide. They are detected by determining the alkalinity of ash and carbonate or bicarbonate by rosalic acid test (Sowmya *et al.*, 2013; Dairy technology, 2017 or http://dairy-technology.blogspot.com/2014/11/neutralizers_16.html).

A. Rosalic acid test for the detection of carbonate and bicarbonate in milk

Take 5 ml of milk in a test tube and add 5 ml ethanol 95% followed by 4-5 drops of rosalic acid. If the colour of milk changes to pinkish red, then it is inferred that the milk is adulterated with sodium carbonate / sodium bicarbonate and hence unfit for human consumption.

This test will be effective only if the neutralizers are present in milk. If the added neutralizers are nullified by the developed acidity, then this test will be negative. In that case, the alkaline condition of the milk for the presence of soda ash has to be estimated.

B. Alkalinity Test

The presence of neutralizers can generally be detected by determining the alkalinity of ash. A known quality of milk is heated and converted into ash.

Take 20 ml of milk in a silica crucible and then the water is evaporated and the contents are burnt in a muffle furnace at 550^0C for 1 hour. Cool the basin. Add 10 ml distilled water and mix the contents with a glass rod. Titrate the ash solution using standard O.1N HCl in the presence of 4-5 drops of phenolphthalein indicator solution. Note the volume of HCl solution used till a pink colour is obtained. If the titre value exceeds 1.2 ml, then it is construed that the milk is *adulterated* with neutralizers.

9) Sodium chloride /salt

Five ml of silver nitrate (0.8%) is taken in a test tube and added with 2 to 3 drops of 1% potassium dichromate and 1 ml of milk and thoroughly mixed. If the contents of the test tube turn yellow in colour, then milk contains salt in it. If it is chocolate coloured, then the milk is free from salt (Sharma *et al.*, 2012a).

10) Detection of Ammonium sulphate

Measure out 2ml of milk into the test tube. Add 500ul of 2% sodium hydroxide into the test tube. Mix the contents in the test tube well. Add 500 ul of 2% sodium hypochlorite into the test tube and mix well. Finally add 500ul of 5% phenol into the test tube. Mix the contents thoroughly. Keep the test tubes in a boiling water bath for 20 seconds. Presence of Ammonium sulphate is indicated by the formation of deep blue color in the milk after taking out from the water bath (Kumar *et al.*, 2002).

11) Test for detection of hydrogen peroxide

Take 5 ml milk in a test tube and then add 5 drops of paraphenylene diamine and shake it well. Change of the colour of milk to blue confirms that the milk is added with hydrogen peroxide (Arvind *et al.*, 2012); (Kamthania *et al.*, 2014); (Sharma *et al.*, 2012a).

12) Detection of skim milk powder in milk

If the addition of nitric acid drop by drop in to the test milk sample results in the development of orange colour, it indicates the milk is adulterated with skim milk powder. Samples without skim milk powder shows yellow colour.

13) Detection of vegetable oil in milk

The characteristic feature of milk is its fatty acid composition, which mainly consists of short chain fatty acids such as butyric, caproic, caprylic acid; whereas the vegetable fats consist mainly of long chain fatty acids and hence *adulteration* of vegetable oil in *milk* can be easily found out by analyzing the fatty acid profile by gas chromatography.

14) Melamine detection methods

A combination of results of Kjeldahl and spectrophotometric methods have been used to screen milk adulterated with melamine (Finete *et al.*, 2013). Recently, several methods have been established to determine melamine content. Unfortunately, most of these methods require complicated pre-condition and costly instruments, and are time-consuming. Hence biochemical detection methods are not efficient to identify for melamine contaminated milk especially at low quantity level.

2.1.2 Quantitative (Biotechnological) detection methods

Ensuring an acceptable level of food quality and safety is absolutely necessary to provide adequate protection for consumers and to facilitate trade. There must be a strong network of efficient quality assurance programme to monitor the quality and safety of these foods before reaching to the consumers. This can be largely possible with the application of recent developed biotechnological tools. The use of modern biotechnology should prove to be rapid, sensitive and accurate methods for detection and analysis of adulterants. However, the recent baby formula milk powder contamination incidents have shown that the standards in milk quality control are insufficient in identifying "manipulated" poor-quality milk (Raut, 2016). Hence, different biotechnology options have been released in different research outputs and some are trying to be applied. However, such type of detection methods are mostly practiced in developed rather than developing countries. Since such type of detection techniques needs high investments. Type of quantitative detection techniques depend on the nature of adulterants in milk.

1. LC (Liquid Chromatography) and ELISA (Enzyme Linked Immunosorbent Assay) are the most common techniques used to detect foreign protein (Dai *et al.*, 2013). MIR (Medium Infra Red) spectroscopy performed better to detect the adulterants such as tap water, whey, hydrogen peroxide, synthetic urea and urine (Santos *et al.*, 2013). Like urea, synthetic urine is also used in milk to increase the nitrogen content. In a case study performed in Brazil, urine was detected in all samples of UHT milk by following a chemometric approach (Souza *et al.*, 2011).

2. PCR (Polymerase Chain Reaction) and PAGE (Polyacrylamide Gel Electrophoresis) are usually used to detect milk from different species as adulterants. Cow milk adulteration in caprine milk has been quantified by HPLC/ESI-MS (High Performance Liquid Chromatography/Electrospray Ionization- Mass Spectroscopy) (Chen *et al.*, 2004). This method identifies molecular masses to differentiate between proteins in the milk of cow and goat. Detection of addition of cow milk to goat and ewe milk has been described (Song, Xue, & Han, 2011).

3. On-line liquid chromatography-gas chromatography (LC-GC) has been applied to the detection of vegetable oils in milk fat using β-sitosterol as marker (Willibald *et al.*, 2002).

4. Quantitatively melamine detection has been possible using SERS (Surface Enhanced Raman Spectroscopy) (Zhang *et al.*, 2010). A portable sensor based on SERS has been also developed to detect melamine instantly (Kim *et al.*, 2012). SB-ATR FTIR (Single Bounce Attenuated Total Reflectance - Fourier transform infrared spectroscopy) has been used to quantify melamine in both liquid and powder milk (Jawaid *et al.*, 2013). Different types of mass spectroscopy have been employed to detect melamine in milk products including LC-MS/MS, APCI-MS (Atmospheric Pressure Chemical Ionization-Mass Spectroscopy) and EESI-MS (Extractive Electrospray Ionization Mass Spectrometry) (Yang *et al.*, 2009; Zhu *et al.*, 2009). LC-MS/MS has been used frequently to detect melamine in milk and variety of infant formulas (Chang & Arora 2008; Guelph, 2008; Sherri *et al.*, 2008). HPLC is another choice of technique to quantify melamine in milk and dairy products (Venkatasami and Sowa, 2010). Using Raman band, melamine in dried milk powder has been immediately detected without extracting melamine from milk (Okazaki *et al.*, 2009). A portable screening system based on Laser Raman spectroscopy has been developed to quantify melamine (Cheng *et al.*, 2010).

5. The researchers used Illumina's Solexa sequencing technology to screen for microRNAs in healthy cow milk. They identified 245 highly expressed microRNAs, seven (miR-26a, miR-26b, miR-200c, miR-21, miR-30d, miR-99a and miR-148a) of which show little changes in expression level during the different stages of lactation (Chen *et al.*, 2010).

The researchers believe that these microRNAs can serve as biomarkers for reflecting the quality of milk and dairy products, including milk powder. Manipulated milk would not be able to replicate the expression levels of these microRNAs in a standard test (Chen *et al.*, 2010).

However, mostly such types of detection methods are not accessible at all level of farmer, milk collection points, or at the spot of dairy cooperatives. To this end, the frauders have escaped condemnation and it makes the less effectiveness of such detection techniques to decrease adulteration (Garcia *et al.*, 2012).

3. MILK ADULTERANTS AND RELATED HEALTH COMPLICATIONS

Now a day in developed and developing countries for economical purposes milk is adulterated with various chemicals. This is very dangerous to human life. Milk is most commonly diluted with water. This not only reduces its nutritional value, but contaminated water can also cause additional health problems. The other adulterants used are mainly detergent, foreign fat, starch, sodium hydroxide (caustic soda), sugar, urea, water, salt, benzoic acid, sodium carbonate, formalin, ammonium sulphate and melamine. Unfortunately, some of the adulterants have immediate severe health impact, sometimes other in the long run (Moore *et al.*, 2012).

The detergent in milk can cause food poisoning and other gastrointestinal complications. Its high alkaline level can also damage body tissue and destroy proteins. Both hydrogen peroxides and detergents in milk can cause gastro-intestinal complications, which can lead to gastritis and inflammation of the intestine. While the immediate effect of drinking milk adulterated with urea, caustic soda and formalin is gastroenteritis, the long-term effects are far more serious. Formalin in milk has long-term effects of kidney failure. Urea in milk overburdens the kidneys as they have to filter out more urea content from the body (Kandpa, *et al.*, 2012). Excessive starch in the milk can cause diarrhea due to the effects of undigested starch in colon, however, accumulated starch in the body may prove very fatal for diabetic patients (Singuluri & Sukumaran, 2014).

In addition, carbonate and bicarbonates might cause disruption in hormone signaling that regulate development and reproduction (Manual of Methods of Analysis of Foods: Milk and Milk Products, 2005).

The ingestion of melamine at levels above the safety limit can induce renal failure and death in infants (Domingo *et al.*, 2014; Cheng *et al.*, 2010). Melamine alone causes bladder stones in animal tests. When combined with cyanuric acid, which may also be present in melamine powder, melamine can form crystals that can give rise to kidney stones. These small crystals can also block the small tubes in the kidney potentially stopping the production of urine, causing kidney failure, heart diseases and even death. Melamine has also been shown to have carcinogenic effects in animals in certain circumstances, but there is insufficient evidence to make a judgment on carcinogenic risk in humans.

4. CAUSE FOR MILK ADULTERATION

Milk is one of the best sources for protein, fat, carbohydrate, vitamin and minerals. Unfortunately, milk is being very easily adulterated throughout the world. Possible reasons behind it may include- demand and supply gap, perishable nature of milk, and lack of suitable detection tests (Kamthania et al. 2014).

5. OPTIONS TO MINIMIZE ADULTRATED MILK PRODUCTS

The cause root for milk adulteration were mentioned above. Hence the solution to maintain quality products should starts from the ground that derive milk producer for adulteration.

5.1 Use of easy milk tester at the spot

5.1.1. Machine detection methods

One of the aggravating cause for milk adultration was unavailability of easy, rapid and affordable milk tester machine at the milk supplying. Hence, milk adulteration detection techniques need to be very specific and rapid, and accessible because of the defrauders have escaped condemnation claiming and it makes less effectiveness of the conventional detection techniques (Garcia, Sanvido, Saraiva, Zacca, Cosso, & Eberlin, 2012). This problem is more acute in the developing and under developed countries due to lack of adequate monitoring and law enforcement. Existing common detection techniques are not always convenient and accessible in these countries making it difficult to address the diverse ways of fraudulent adulteration in milk.

Milk analyzer machine are practical option to test adulteration at the spot. These technologies used to measure milk quality even at household levels. They are simple and rapid test procedures. Such types of technologies are used in different developed and developing countries.

There are different types of machine which are perfect tool to identify adulteration on site. They are rapid milk analyzers used to measure milk fat, solids non-fat (SNF), density, protein, lactose, salts, water content percentages, temperature, freezing pointes, PH, conductivity, as well as total solids of sample directly after milking, at collecting and processing. There are also high sensitivity adulteration test stripes for detection of neutralizers, hydrogen peroxide and urea adulteration in raw milk, which work in a very efficient and reliable way all over the world. The price of such test machine ranges from 100 to 1500 US$.

However, these types of machine do not detect melamine. Melamine content in milk is delected by other handheld sensor device. Detecting melamine in milk has become extremely easy, quick and inexpensive. The detector was developed by researchers in Indian Institute of Science (IISc), Bangalore. The sensore has a very sensitivity as it can detect melamine even at low concentration level of 0.5 ppm in raw milk.

5.1.2. Ready to use kit

One of the technologies used to ensure quality of milk at the spot is a biochemical method which is fast and easy using commercialized ready-to-use kit for detection of commonly used adulterants in milk. This kit can detect the presence of all adulterant except melamine, by comparing the colors developed after addition of test reagents to milk. The Kits have manual and hence can be used by unskilled persons with little or no training at all.
The kit is suited for specific usage, at household level, dairy cooperative, society level and dairy plant level.

5.2. IMPROVE PROFIT MARGIN OF DAIRY FARMERS

Approximately 150 million farm households around the globe are engaged in milk production. In most developing countries, milk is produced by smallholders and it contributes to household food security. In recent decades, developing countries have increased their share in global dairy production. In this case demand of milk is increasing, however, reversely there is shortage in supply (FAO, 2010). To this end, milk is being very easily adulterated throughout the world to fill the demand and supply gap (Faraz *et al.*, 2013). Augmenting cow's milk production is a prerequisite for improving the profitability of dairy farmers and to decrease consequence of producing adulterated milk. In addition to this even in high production season milk is adulterated by producer. Since all of the milk is not sale on one day, they use hazardous chemical to preserve the milk to sale it on the other day. Hence to maintain quality product, and then to increase profit margin of the dairy farmers there are different biotechnological and technical options that have been practiced in developed and developing countries. In general, biotechnology brings production and quality revolution together.

5.2.1. Improving Milk Production by Biotechnological Application

Biotechnology can broadly be defined as the technology by which one can produce useful products from raw materials with help of living organisms or other biological processes

Recent development in biotechnology have open up new and exciting possibilities in dairying for enabling the availabilities of milk and milk products within the reach of poor and to cater the need of large section of population. Thus, biotechnological interventions can not only improve milk production and quality but it can benefit through improving commercial value of milk for local consumption as well as export market through milk processing and longer shelf life (Madan, 2005; Singh & Gandhi, 2015). Biotechnology has also made a significant impact on the dairy industry by improving the quality of animals through application of assisted reproductive techniques such as artificial insemination (AI) and embryo transfer (ET). Based on a large dataset study in Canada indicates that for each 1000 kg increase in mean milk production was due to 0.7 % increase in pregnancy rate (LeBlanc, 2010).

5.2.1.1 Improving milk production through reproductive biotechnology

1. **Artificial Insemination (AI)**: is the most widely used biological technology in developed and developing nations in livestock farming (Cardellino *et al.*, 2003). It indirectly improves milk production through improving the quality of dairy animals. Its notable benefits include

the extension use of males in desired traits over a vast number of females, which would be unachievable through natural service and the prevention of transmittable venereal diseases. Frozen semen allows genetic progress to be disseminated worldwide. Today, artificial insemination is applied in other animals (Foote, 2002). However, it has been most extensively applied in cattle production, particularly associated with dairy cattle.

2. **Embryo Transfer (ET)**: Many industries, including the dairy industry, started using embryo transfer to enhance the quality of their products. However, due to the fact that cattle typically produce one embryo at a time, this technique remained largely limited until follicle-stimulating hormone (FSH) was discovered in the 1950s. The discovery of FSH greatly improved the prospects of embryo transfer as production of multiple eggs increases the success rate of such techniques. FSH is a chemical that makes the donor produce several mature eggs, each of which is capable of being fertilized and producing an offspring. In addition, this hormone helps to have planned calving through a technique of oestrus synchronisation and superovulation. To this end, multiple ovulations and the freezing of embryos enables the extended use of superior female genetic material, and whole genomes (diploid) across continents at low transportation costs.

Embryo transfer has several advantages.
- Producing embryos (and subsequently calves) with desired qualities in a short time
- Allowing birth to an increased number of high quality offspring for dairying (calves that provide better quality milk and other dairy products)
- Produce high quality offspring from low quality cattle,
- Preventing diseases in cattle.
- Moreover, excess embryos can also be retained for long periods by freezing.

Most developing countries have limited financial resources. In addition, equipment and supplies tend to be more expensive than they are in developed countries due to transportation costs, import tariffs, lack of hard currency etc. In general, the inappropriateness of ET for developing countries is ascribed to lack of infrastructure. These make ET technology prohibitive in these countries. Seidel and Seidel (1992) pointed out that a lot of equipment associated with ET is not essential for successful utilization of the technology. For example, freezing equipment are no more effective than dry ice or alcohol baths for freezing embryos as these save labour and are more convenient. Thus, by combining good imagination with knowledge of basic principles, the technology can be successfully adapted to local conditions in developed countries.

To this end, biotechnological research is important in order to respond to the pressure of producing more food from animals then to cater food requirement of the ever-growing human population (Kahi and Rewe, 2008).

5.2.1.2 Improve milk quality through processing technology

Milk is the most perishable product in the food items. Spoilage of milk is one of the driving forces for milk adulteration. Since, to increase the shelf life of milk, different adulterants are added to it. Hence, processing of milk in to different value-added product will minimize milk lose due to spoilage and in turn it maximizes profit margin in safe way rather than expanding profit through adulteration.

Milk processing has advantage for producer since:
- Such technologies can create better market access through supplying different products with the need of consumer.
- The technique improves quality and safety of milk. In turn it improves the profit margin of the producer than selling fresh milk.
- Processed product can be stored for longer periods without being spoilt so that it provides regular income for producer.
- Milk with good keeping quality fetches good market prices. Milk that is sour or otherwise unpalatable cannot be sold for direct use. At this time instead of selling raw milk, processed milk products like butter, yogurt or cheese, have better opportunity to earn more income and also
- It prevents spoiling of extra or unsold milk especially in high production season.

Processing technology will be one of the options to protect milk adulteration in the supply chain (starts from producer until it reaches to trader) through extending milk shelf life. Accessibility and/or affordability of the milk processing equipment at all level of milk collection site or at farmer levels limits the use of the technologies. Hence, options should be identified to maintain improved quality product with adopting simple milk processing equipment and technique which is financial viable as well as accessible to the producer. Processing equipment should be suitable for small-scale producers, especially for developing country, to produce simple value-added milk products such as yogurt, butter and cheese.

5.2.2 Labeling of Milk quality (*as A1 and A2* milk)

Producing different type of milk is a pivotal in increasing profit margin of the farmer through getting income from various sources. Milk type is depending on the breed of cow it came from. Dairy cow's milk around the world are classified in to A1 and A2 milk type based on the type of protein involved in their milk. Hence the dairy farmer can maximize their income through labeling the whole milk as A1 and A2 milk.

What are A1 and A2 milk?

Milk from dairy cows has been regarded as nature's perfect food, providing an important source of nutrients including high quality proteins, carbohydrates and selected micronutrients. More than 95% of the cow milk proteins are constituted by caseins and whey proteins. Among the caseins, beta casein is the second most abundant protein and has excellent nutritional balance of amino acids (Healthline, 2015).

Different mutations in bovine beta casein gene have led to 12 genetic variants and out of these A1 and A2 are the most common. The A1 and A2 variants of beta casein differ at one amino acid position. Hence, milk with variant A1 of β -casein named as A1 milk and the other with A2 β casein as A2 milk.

Gastrointestinal proteolytic digestion of A1 variant of β-casein leads to generation of bioactive peptide, beta casomorphin 7 (BCM7) (IvanoDe and Stefano, 2010; Boutrou *et al.*, 2013; Handpicked, 2013). Which has been shown to alter gastrointestinal function (slowing down bowel movements from stomach to anus) and increase inflammation in the gut. Milk with variant A1 of beta-casein has BCM-7 level four-fold higher than in A2 milk. Hence, scientists believe that A2 alleles are a source for safe milk (Elliott *et al.*, 1999; Mishra *et al.*, 2009).

A1 β-casein is common in dairy cows of north European origin such as Friesian, Ayrshire, British Shorthorn, and Holstein. Predominantly A2 β-casein is found in the milk of Jersey, in Southern French breeds, Charolais and Limousin and in the Zebu original cattle of Africa (Ng-Kwi-Hang & Grosclaude, 1992). Regular milk contains both A1 and A2 beta-casein, but A2 milk contains only A2 beta-casein. Hence, the health effects of milk may depend on the breed of cow it came from. An ingenious hypothesis about A1/A2 milk was developed by RB Elliott and CNS McLachlan and collaborators in the 1990s. Following their hypothesis, different studies indicate that A1 beta-casein may be harmful, and that A2 beta-casein is a safer choice.

As a result of this, currently A2 milk is being marketed as a healthier choice than regular milk. It is claimed that A1 have several health benefits and to be easier to digest for people who are lactose intolerant (IvanoDe and Stefano, 2010; Boutrou *et al.*, 2013; Healthline, 2015).

Health related scientific evidence in A1/A2 milk

In addition to this, epidemiological evidences claim that consumption of beta-casein in A1 milk is associated with a risk factor for type-1 diabetes, coronary heart disease, arteriosclerosis, sudden infant death syndrome, autism, schizophrenia etc.(Laugesen and Elliott, 2003; Tailford *et al.*,2003; IvanoDe and Stefano, 2010; Boutrou *et al.*, 2013). However, European Food Safety Authority review of scientific literature found that there was insufficient evidence to prove that bioactive peptides in A1 milk have a negative effect on health (EFSA, 2009). This has been the reason for the continuity of "A1 vs A2" debate.

Review of the scientific evidence in relation to A1 milk consumption and related health effect:
1. Several studies indicate that drinking of A1 milk during childhood may increase the risk of type 1 diabetes (Elliott *et al.*, 1999; Mclachlan, 2001; Laugesen and Elliott, 2003; Birgisdottir *et al.*, 2013).

2. Observational studies have linked the consumption of A1 milk with an increased risk of heart disease (Mclachlan, 2001; Laugesen and Elliott, 2003; Truswell, 2005). However other studies (Chin *et al.*, 2006; Venn *et al.*, 2006) disproof this result that the consumption of A1 milk has no significant adverse effect compare to A2 on risk factor for heart diseases.

3. Suden Infant Death Syndrem (SIDS) is defined as the unexpected death of an infant, without an apparent cause. It is the most common cause of death in infants less than one year of age. Some researchers reported that BCM-7 may be involved in some cases of SIDS (Wasilewska *et al.*, 2011).

4. Autism is a mental condition characterized by poor social interaction and repetitive behavior. In theory, peptides like BCM-7 might play a role in the development of autism and strongly associated with an impaired ability to plan and perform actions. Studies indicated that drinking cow's milk may worsen behavioral symptoms in autistic children (Lucarelli *et al.*,1995; Kost *et al.*, 2009). However, some studies do not support all of the proposed mechanisms (Hunter *et al.*,2003; Cass *et al.*, 2008; Navarro *et al.*, 2015; Millward *et al.* 2008).

5. Digestive Health- Lactose intolerance is defined as the inability to fully digest the sugar (lactose) found in milk. This is a common cause of bloating, gas and diarrhea. Different research trial showed that A1 milk may cause softer stools (Hos, 2014). Studies in rodents indicate that A1 beta-casein may significantly increase inflammation in the digestive system (Ui *et al.*, 2014; Barnette, 2014).

Hence, due to different research report the A1/A2 debate is still up in the air. However, there are many A2 dairy industries are developing in different countries. The A2 milk company continues to develop a portfolio of intellectual property, including trademarks and patents. The A2 milk company also invests in targeted research and development associated with the benefits of the A2 protein. Moreover, Following the 2008 Chinese milk scandal, China has been dramatically increasing importation of A2 milk (Jared, 2015). Sales of A2 milk in Australia and New Zealand boosted significantly following 2007, after the publication of a book, *Devil in the Milk* by Keith Woodford, about A1 beta-casein and its perceived dangers to health (A2 Corporation Press, 2008). Moreover, the A2 milk company launched an infant formula in New Zealand and

Australia in September 2013 (Adams, 2013). A2 milk is more preferable by consumer. Moreover, different companies also incorporate A2 milk protein in their food products as one ingredient (Hawthone, 2014).

Establishment of A2 milk industry even will facilitate the exportation of such type of milk to different countries in the world. Hence, countries that have A2 (like Zebu) cattle has an opportunity of labeling their milk as A2 and it support the producer to maximize their profit margin through providing more choice of dairy products to the consumer, either A1 or A2 milk.

What do consider here?

- *The Bos Indicus cow is the African native breed that produces the A2 milk with the good quality protein*
- *But it has been conveniently replaced by the high-yielding cross breed, popularly known as Holstein Friesian which provides the A1 variety of milk. This is something we need to consider beyond the 'type' of milk."*
- *Hence, we have not neglected our native breeds whose milk was always considered to be superior even with health benefits. Hence we have to give recognition for real exploitation of their potential*

6. CONCLUSIONS AND RECOMMENDATIONS

Dairy farmers or traders deliberately adulterate the milk with different products. Commonly, economically motivated adulterant like water is added to milk to increase volume, however, which decreases specific gravity of milk. Then to retain the gravity, shelf life, and color of the milk hazardous chemicals like urea, formalin, detergents, ammonium sulphate, boric acid, caustic soda, benzoic acid, salicylic acid, hydrogen peroxide, sugars and melamine are added.

Hence to test the milk against to adulteration, qualitative and quantitative detection methods are used. Milk adulteration is more acute in the developing and under developed countries due to lack of adequate monitoring and law enforcement. In addition, existing common detection techniques are not always convenient and accessible in these countries making it difficult to address the diverse ways of fraudulent adulteration in milk. Hence, this calls for combined efforts from scientific communities and the regulatory authorities through the development, implementation and dissemination of better techniques for the detection of milk adulteration. Furthermore, simple detection methods like simple milk tester machine, mobile testing lab or testing kit should be stationed at various milk sales outlets. In addition, there must be clear rules and regulations to penalize the criminals. Creating awareness and access to information on type of milk adulterants and related health risks can plays vital role in these countries to overcome this issue.

Demand and supply gup is the main motor to milk adulteration. The other causes are big need to higher financial gain, and perishable nature of the milk also aggravates this case. Hence, to minimize milk adulteration and to maintain quality product the above driving problems should get solution. Reproductive and processing biotechnology are practical options in improving production and efficiently use the produced product without spoilage, respectively. In addition, these are feasible solution to maintain quality product and to ensure the increased profit margin of milk producers by increasing volume as well as improving commercial value of milk. However, all available biotechnologies are not appropriate or relevant to all countries. Developing countries may need to adapt some of these technologies before they can use them.

Moreover, researchers need to be exposed to new technologies or procedures to appreciate the power and limitations of such technologies in such countries.

Increasing option of value added products through processing and/or labeling (A1 or A2) type of milk will maximize the profit of the actors also. Such techniques can create better market access through supplying different products with the need of consumer. However, information gap limits the use of the techniques. Hence, options should be identified for adopting simple technique which helps in diversifying income from the already produced milk in turn to minimize milk adulteration.

7. REFERENCES

A2 Corporation Press (2007): A2 Milk Sales in NZ / Australia Increase Substantially A2 Corporation Press Release November, 2007.

Adams, Christopher (2013): "New A2 infant formula ready for China". The New Zealand Herald. Auckland, press 22 April 2013.

Arvind Singh GC, Aggarwal A, Kumar P. (2012): Adulteration Detection in Milk. Res News For U (RNFU). 2012;5:52–5.

Bansal A and Singhal OP. (1991): Preservation of milk samples with formalin- Effect on Acidity. Indian Journal of Dairy Sciences, 44: 573.

Boutrou R, Gaudichon C, Dupont D, Jardin J, Airinei G, Marsset-Baglieri A, Benamouzig R, Tomé D, Leonil J. (2013): Sequential release of milk protein-derived bioactive peptides in the jejunum in healthy humans. Am J Clin Nutr. 2013 Jun; 97 (6):1314-23. doi: 10.3945/ajcn.112.055202. Epub 2013 Apr 10.

Chang E, and Arora I. (2008): Simultaneous, Fast Analysis of Melamine, Cyanuric Acid, and Related Compounds in Milk and Infant Formula by LC/MS/MS. 2008. (pp. 1- 4): http://www.ingenieria-analitica.com/downloads/dl/file/id/2737/product/

Chen RK, Chang LW, Chung YY, Lee MH, Ling YC. (2004): Quantification of cow milk adulteration in goat milk using high-performance liquid chromatography with electrospray ionization mass spectrometry. Rapid Commun Mass Spectrom. 2004;18:1167–71.

Chen, X. Gao C, Li H, Huang L, Sun Q, Dong Y, Tian C, Gao S, Dong H, Guan D, Hu X, Zhao S, Li L, Zhu L, Yan Q, Zhang J, Zen K, Zhang CY (2010): Identification and characterization of microRNAs in raw milk during different periods of lactation, commercial fluid, and powdered milk products. Cell Res. doi:10.1038/cr.2010.80 (2010).

Cheng Y, Dong Y, Wu J, Yang X, Bai H, Zheng H. (2010): Screening melamine adulterant in milk powder with laser Raman spectrometry. J Food Composit Anal. 2010;23(2):199–202.

Chikkappa Udagani & Thimmasandra Narayan Ramesh, (2014): Quantitative Detection of Melamine Adulteration in Milk Powder: Using G-ray Spectroscopic Technique. Impact: International Journal of Research in Applied, Natural and Social Sciences (Impact: Ijranss). ISSN (e): 2321-8851; ISSN (p): 2347-4580. vol. 2, Issue 2, Feb 2014, 81-88

Dai X, Zhao Y, Li M, Fang X, Li X, Li H, Xu B. (2013): Determination of urea in milk by liquid chromatography-isotope dilution mass spectrometry. Analytical Letters. 2013;45:1557–65.

Domingo E, Tirelli AA, Nunes CA, Guerreiro MC, Pinto SM. (2014): Melamine detection in milk using vibrational spectroscopy and chemometrics analysis: A review. Food Res Int. 2014;60:131–9.

EFSA, (2009): Review of the potential health impact of β-casomorphins and related peptides. European Food Safety Authority (EFSA) Journal. **7** (2): 231r. doi:10.2903/j.efsa.2009.231r.

Elliott RB, Harris D P, Hill JP, Bibby NJ, Wasmuth HE. (1999): Type I (insulin-dependent) diabetes mellitus and cow milk: casein variant consumption. Diabetologia. 1999;42:292-6. [PubMed]

Ellis DI, Brewster VL, Dunn WB, Allwood JW, Golovanov AP, Goodacre R.(2012): Fingerprinting food: current technologies for the detection of food adulteration and contamination. Chem Soc Rev. 2012;41(17):5706–27.

Embrapa (2014): Estatísticas de produção de leite (Statistics of milk production). Brasília: Brazilian Agriculture Research Corporation. Available in. Accessed in June 2014

Encyclopedia of Dairy Sciences. (2011): 2011;2:887–97.

FAO (FOOD AND AGRICULTURE ORGANIZATION OF THE UNITED NATIONS). (2010): Pro-Poor Livestock Policy Initiative: Status and Prospects for Smallholder Milk Production.A Global Perspective.FAO. Rome, 2010.

Faraz A, Laleef M, Mustaf MI, Akhtar P, Yagoob M, (2013): Detection of adulteration, chemical composition and hygienic status of milk supplied to various canteens of educaational institutes and public places in Fisal Abad. J Anim Plant Sci 23: 119-124.

Finete V M, Gouvea M M, Marques F C and Netto A D P. (2013): Is it possible to screen for milk or whey protein adulteration with melamine, urea and ammonium sulphate, combining Kjeldahl and classical spectrophotometric methods? Food chemistry 141,3649- 3655

Fischer W, Schilter B, Tritscher A, Stadler R. (2011): Contaminants of milk and dairy products: contamination resulting from farm and dairy practices.

Garcia JS, Sanvido GB, Saraiva SA, Zacca JJ, Cosso RG, Eberlin MN. (2012): Bovine milk powder adulteration with vegetable oils or fats revealed by MALDI-QTOF MS. Food Chem. 2012;131(2012):722–6. 131 722–726.

Guelph U. (2008): Determination of residues of melamine and cyanuric acid in animal food by LCMS/MS. Guelph: University of Guelph, Laboratory Services Division; 2008:1-31.

Handpicked Press (2013): What is A1 Versus A2 milk? January,2013. Newsletter. http://handpickednation.com/

Hawthone, M. (2014): "This means war in a milky way", *The Age*, Melbourne, p. 8, retrieved 27 June 2014.

Healthline Press. (2015): A1 vs A2 Milk - Does It Matter? Health line press Written by Atli Arnarson, January, 2015. https://www.healthline.com/nutrition/a1-vs-a2-milk

Henno M., M. Ots, I. Jõudu, T. Kaart, Kärt O. (2008): "Factors affecting the freezing point stability of milk from individual cows", International Dairy Journal, vol. 18, 2008: 210-215.

Ilie l.i., L. Tudor, A. Maria G. (2010): Research concerning the main factors that influence the freezing point value of milk. Lucrări stiinlifice medicină veterinară vol. Xliii (2), 2010, timisoara 228

Ingelfinger JR. (2008). Melamine and the global implications of food contamination. New England J Med. 2008;359(26):2745–8.

IvanoDe N. and Stefano C. (2010): Occurrence of β-casomorphins 5 and 7 in commercial dairy products and in their digests following *in vitro* simulated gastro-intestinal digestion. Food Chemistry. Volume 119, Issue 2, 15 March 2010, Pages 560-566.https://doi.org/10.1016/j.foodchem.2009.06.058Get rights and content

Jared Lynch , (2015): "A2 weighs up building baby formula plant to cash in on China demand". Sydney Morning Herald. Retrieved 22 December 2015.

Jawaid S, Talpur FN, Sherazi ST, Nizamani SM, Khaskheli AA. (2013): Rapid detection of melamine adulteration in dairy milk by SB-ATR–Fourier transform infrared spectroscopy. Food Chem. 2013;141(3):3066–71.

Kahi A.K. and Rewe T.O. (2008): Biotechnology in livestock production: Overview of possibilities for Africa. African J. of Biotech 7: 4984-4991.

Kamthania M, Saxena J. Saxena K. and Sharma D.K. (2014): Milk Adulteration: Methods of Detection &Remedial Measures. National Conference on Synergetic Trends in engineering and

Technology (STET-2014) International Journal of Engineering and Technical Research. 2014;1:15–20.

Kandpal SD, Srivastava AK, Negi KS. (2012): Estimation of quality of raw milk (open & branded) by milk adulteration testing kit. Indian J Comminity Health. 2012;24:3.

Khan KM, Krishna H, Majumder SK, Gupta PK. (2014): Detection of urea adulteration in milk using near-infrared Raman spectroscopy. Food Anal. Methods. 2014

Kim A, Barcelo SJ, Williams RS, Li Z. (2012): Melamine sensing in milk products by using surface enhanced Raman scattering. Anal Chem. 2012;84(21):9303–9.

Kumar R, Singh DK, Chawla NK. (1998): Adulteration/contamination of milk demystified. Indian Dairyman. 1998;50:25–33.

Laugesen M, and Elliott R. (2003): Ischaemic heart disease, type 1 diabetes, and cow milk A1 beta-casein. N Z Med J. 2003;116:U295. [PubMed]

Lin M, He L, Awika J, Yang L, Ledoux DR, Li HA, Mustapha A. (2008): Detection of melamine in gluten, chicken feed, and processed foods using surface enhanced Raman spectroscopy and HPLC. J Food Sci. 2008;73(8):T129–34.

Liu B, Lin M, Li H. (2010): Potential of SERS for rapid detection of melamine and cyanuric acid extracted from milk. Sensing Instrum Food Qual Saf. 2010;4(1):13–9.

McLachlan CN. (2001): beta-casein A1, ischaemic heart disease mortality, and other illnesses. Med Hypotheses. 2001 Feb;56(2):262-72.

Madan M.L. (2005): Animal biotechnology: applications and economic implications in developing countries. *Rev. Sci.*

Manual of Methods of Analysis of Foods (2005): Milk and Milk Products. D. G. o. H. Services (Ed.): Ministry of Health and Family Welfare, Government of India

Maryam J .A (2017): Review Paper on Melamine in Milk and Dairy Products. Dairy and Vet Sci J. 2017; 1(4): 555566.DOI: 10.19080/

Mishra BP, Mukesh M, Prakash B, Monika S, Kapila R, Kishore A., (2009): Status of milk protein, b-casein variants among Indian milk animals. Indian J Anim Sci. 2009;79:72–5.

Moore JC, Spink J, Lipp M. (2012): Development and Application of a Database of Food Ingredient Fraud and Economically Motivated Adulteration from 1980 to 2010. J Food Sci. 2012;77:R108–16.

Noel C., (2005): The fight against food adulteration. The royal society of chemistry. London,

Okazaki S, Hiramatsu M, Gonmori K, Suzuki O, Tu AT. (2009): Rapid nondestructive screening for melamine in dried milk by Raman spectroscopy. Forensic Toxicol. 2009;27:94–7.

Panesar PS (2011): Fermented Dairy Products: Starter Cultures and Potential Nutritional Benefits. Food Nutr. Sci.2(1):47-51.

PerkinElmer, (2013): Food Fraud: Milk Adulteration with Melamine – Screening, Testing and Real-Time Detection. PerkinElmer for better Inc. 940 Winter Street Waltham, MA 02451 USA

Raut V.S. , (2016): Milk Adulteration and Detection. A Review sensor letters Vol. 14, 4-18.

Renny EF, Daniel DK, Krastanov AI, Zachariah CA, Elizabeth R. (2014). Enzyme Based Sensor for Detection of Urea in Milk. Biotechnol Biotechnol Equip 19: 198-201

Santos PM, Pereira-Filho E R and Rodriguez-Saona L.E.(2013): application of hand-held and portable infrared spectrometers in bovine milk analysis .J.Agric.FoodChem. 61(2013)1205-12011

Saxelin M, TynkkynenS, Mattila-Sandholm T, de Vos W (2005): Probiotic and other functional microbes: from markets to mechanisms. Curr. Opin. Biotechnol.16:204–211

Sharma R., Rajput Y. S., Barui A. K., & N., L. N. (2012)a: Detection of adulterants in milk, A laboratory manual. In N. D. R. Institute (Ed.). Karnal-132001, Haryana, India. 2012.

Sharma R, Sanodiya BS, Bagrodia D, Pandey M, Sharma A, Bisen PS. (2012)b: Efficacy and Potential of Lactic Acid Bacteria Modulating Human Health. Int. J. Pharma Bio Sci. 3(4): 935-948.

Sherri Turnipseed, CC Cristina Nochetto, David N. Heller. (2008): Determination of melamine and cyanuric acid residues in infant formula using LCMS/MS. In: U.S. Food nd Drug Administration. 2008

Singh P, and Gandhi N. (2015): Milk preservatives and adulterants: processing, regulatory and safety issues. Food Rev Int. 2015;31(3):236–61.

Singhal OP. (1980): Adulterants and methods for detection. Indian Dairyman. 1980;32(10):771–4.

Singuluri H, and Sukumaran M. (2014): Milk adulteration in hyderabad, India a comparative study on the levels of different adulterants present in milk. J Chromatograph Separat Techniq. 2014;5:212.

Song H, Xue H, Han Y. (2011): Detection of cow's milk in Shaanxi goat's milk with an ELISA assay. Food Control. 2011;22(6):883–7.

Souza S S, Cruz A G, Walter E H M, Faria J A F, Celeghini R M S, Ferreira M M C. (2011): Monitoring the authenticity of Brazilian UHT milk: A chemometric approach. Food Chemistry, 124, 692–695

Sowmya R.K.P. Indumathi, S. Arora, V. Sharma, and Singh A.K. (2013): Detection of calcium based neutralizers in milk and milk products by AAS journal of food science and tehnology.2015 feb;52(2):1188-1193.

Spink J, and Moyer DC. (2011): 'Defining the public health threat of food fraud. J Food Sci 76: R157-R163.

Tailford KA, Berry CL, Thomas AC, Campbell JH. (2003): A casein variant in cow's milk is atherogenic. Atherosclerosis. 2003; 170:13–19. [PubMed]

Tanzina Azad and Shoeb A. (2016): Common milk adulteration and their detection techniques. International Journal of Food Contamination20163:22 *Tech. Off. Int. Epitz.* 2005, 24(1): 127-139.

Truswell AS. (2005): The A2 milk case: a critical review. Eur J Clin Nutr.May;59(5):623-31.

Venkatasami G, Sowa JR. (2010): A rapid, acetonitrile-free, HPLC method for determination of melamine in infant formula. Analytica Chimica Acta. 2010;665:227–30.

Venn BJ, Skeaff CM, Brown R, Mann JI, Green TJ. (2006): A comparison of the effects of A1 and A2 beta-casein protein variants on blood cholesterol concentrations in New Zealand adults. Atherosclerosis. 2006 Sep;188(1):175-8. Epub 2005 Nov 18.

Wasilewska J, Sienkiewicz-Szłapka E., Kuźbida E., Jarmołowska B., Kaczmarski M., Kostyra E. (2011): The exogenous opioid peptides and DPPIV serum activity in infants with apnoea expressed as apparent life threatening events (ALTE). Neuropeptides. 2011 Jun;45(3):189-95. doi: 10.1016/j.npep.2011.01.005. Epub 2011 Feb 21.

Willibald K., Fabiola D, Claudia H, Hans-Georg S, Karl-Heinz E, (2002): Rapid detection of vegetable oils in milk fat by on-line LC-GC analysis of β-sitosterol as marker Europian journal of lipid science and technology. Volume 104, November 2002. Pages 756–761

Xin H, and Stone R.(2008): Tainted milk scandal. Chinese probe unmasks high-tech adulteration with melamine. Science. 2008;322:1310–1.

Yang S, Ding J, Zheng J, Hu B, Li J, Chen H, (2009): Detection of melamine in milk products by surface desorption atmospheric pressure chemical ionization mass spectrometry. Analytical Chemistry. 2009;81:2426–36.

Zhang XF, Zou MQ, Qi XH, Liu F, Zhu XH, Zhao BH. (2010): Detection of melamine in liquid milk using surface-enhanced Raman scattering spectroscopy. J Raman Spectroscopy. 2010;41(12):1655–60.

Zhu L, Gamez G, Chen H, Chingin K, Zenob R.,(2009): Rapid Detection of melamine in untreated milk and wheat gluten by ultrasound assisted extractive electrospray ionization mass spectrometry (EESI-MS). Chem Comm. 2009;5:559–61.